How to Have a Healthy Body

How to Have a Healthy Body

HERON
BOOKS

Published by
Heron Books, Inc.
20950 SW Rock Creek Road
Sheridan, OR 97378

heronbooks.com

Special thanks to all the teachers and students who
provided feedback instrumental to this edition.

ISBN: 978-0-89-739209-9

Printed in the USA

28 February 2022

At Heron Books, we think learning should be engaging and fun. It should be hands-on and allow students to move at their own pace.

To facilitate this we have created a learning guide that will help any student progress through this book, chapter by chapter, with confidence and interest.

Get learning guides at
heronbooks.com/learningguides.

For teacher resources,
such as a final exam, email
teacherresources@heronbooks.com.

We would love to hear from you!
Email us at *feedback@heronbooks.com.*

IN THIS BOOK

CHAPTER 1

What Is a
Healthy Body?

CHAPTER 1

What Is a Healthy Body?

When your body is **healthy**, it is operating well. It has lots of energy and doesn't get sick very easily.

If your body is healthy, you can use it to do what you want to do.

Think of a bicycle that someone hasn't taken very good care of. It is dirty and rusty in places. The wheels don't go around easily. The pedals slip sometimes and the brakes don't work very well.

You wouldn't want to ride this bicycle, because it would be hard to ride and might break completely. When you do ride it, you would be worried about the bicycle.

A body that is *not* healthy is like that.

Now imagine a bicycle that someone has taken very good care of. It may not be new, but it is clean and works well. The pedals are easy to turn, the brakes work great, and it's fun to ride.

When you ride this bicycle, you don't have to think much about the bicycle. You can think about where you're going and what you're doing.

A healthy body is like that.

In the same way that you can learn to take care of a bicycle so that it runs well, you can learn to take care of your body so that it runs well.

After all, who can take better care of it than you?

LET'S DO THIS!
Healthy or Not?

Steps:

1. a) Think of a bicycle or other machine you are familiar with and list some of the things about it that are "healthy." In other words, what things are right and working well?

   b) Then list some of the things about it that are unhealthy (things that are not right and don't work well).

2. a) List some ways you have seen that a body can be unhealthy.

   b) Then list some of the things about a healthy body that are important to you.

CHAPTER 2
Keeping Clean

CHAPTER 2

Keeping Clean

Being clean, whether it's your body or your house, is important to being healthy. Dirt on your body, or old food in your room or house, gives a good place for germs to grow.

Dirt and old food also give a place for insects to be and grow, and even for rats and mice in a house. Insects, and rats and mice, can carry germs. When germs grow too much, they can cause sickness.

At one time people in many places didn't know that being dirty and having old food around could help cause sickness. They did not take baths very often, and they left old food around or threw it into the street. There were many rats and insects in the cities. Sometimes the people were dirty enough that small insects called lice lived on their bodies.

Sometimes there would be a lot of sickness in these cities, and many, many people would die from these sicknesses. The sicknesses would spread to the whole countryside.

Finally, someone found out that being clean helps prevent some kinds of sickness. Since then, people have been more careful to keep themselves clean. When they do this well, they don't have so much trouble with diseases.

Here are some important things to know about keeping clean:

1. Always wash your hands after you go to the bathroom.

2. Wash your hands before you eat. You should also wash your hands before you make food for yourself or anyone else.

3. Clean your body regularly by taking baths or showers. (People who don't do this are not only dirty, they also smell bad. The smell comes from germs that grow on the body when it isn't kept clean often enough.)

4. Brush your teeth two times a day.

5. Keep the space around you clean. Especially don't leave old food around.

6. Wash dishes very well after you use them. Do not cook with dirty dishes and do not eat from dirty dishes.

7. Wash your clothes after you wear them.

It's not hard to do these things, and it will help you and the people around you stay healthy.

LET'S DO THIS!
Keep Clean

Steps

1. Practice saying in your own words the seven things to know about keeping clean until you are comfortable saying them.

2. Tell another person what they are without reading them.

LET'S DO THIS!
Keep Clean Better

Steps

1. Using the seven steps given in the chapter, think of something you could do better in keeping yourself, your belongings, and your surroundings clean.

2. Start doing it.

3. When you have done it for a while, write down what you did and how it affected you.

4. Show your write-up to another person.

CHAPTER 3

Food

CHAPTER 3

Food

One important way to have a healthy body is to eat the things the body should have. Your body gets what it needs to grow, have energy and be healthy from the food you feed it. You could say that food for your body is like gas for a car. It's what makes your body run. To make it run well, you need to give it good food.

NUTRIENTS

The things in food that your body needs for life and growth are called **nutrients**.

Different kinds of food have different amounts of nutrients. Here are the main kinds of nutrients, along with examples of foods that have the most of that kind.

Protein

Protein (PRO-teen) is what your body uses to build your muscles, bones, teeth, skin and even blood.

Here are some foods that have a lot of protein:

meat   eggs   fish   yogurt

milk   cheese   turkey and chicken

All of these come from animals. It is easiest to get protein from animal foods. If someone doesn't want to eat animal foods, they have to work a bit harder to get enough protein, but here are some choices.

soybeans

rice and beans together

peanut butter and other nuts

Because bodies build with protein, it's especially important for growing bodies to get enough protein.

Carbohydrate

Carbohydrate (CAR-bo-HI-drate) is what your body uses to make quick energy. Shortly after, there can be an energy drop.

Here are some foods that have a lot of carbohydrate:

bread   chips   crackers

potatoes   rice   noodles

spaghetti   cereal   fruit

corn   squash

foods with lots of sugar, like cookies, cake, candy and sodas

When you eat a lot of carbohydrate, your body can't use it all, and makes the extra into fat. The body gets fatter and less healthy. Many carbohydrates, especially those with lots of sugar, also don't provide the body with other useful nutrients.

Fat

Fat is what your body uses to make long-lasting energy.

Here are some foods that have fat your body can use:

butter   meat   cheese

eggs   avocado   nuts

milk   salmon   tuna

There are other kinds of fat and oil that are not as good for your body, but you can learn more about that later.

Vitamins and Minerals

Vitamins and **minerals** are special, tiny nutrients that are found in many of the foods we eat, but sometimes we don't get enough of them and need to add them to our diet (often as pills). The vitamins have alphabet names, like Vitamin A, Vitamin E, Vitamin C. The minerals have names like salt and calcium.

HELPERS

Here are two things that aren't nutrients but that bodies need to be healthy.

Fiber

This is a different kind of food, because it helps the body, but the body does not use it up. **Fibers** are tiny "strings" in foods like carrots, celery, and oatmeal. If you chew on a celery stick, you may notice the long strings of fiber. Fiber helps the body do its work of moving the food through the stomach and intestines so it can be broken down and used by the rest of the body. The fiber itself doesn't get digested—it just goes on through!

Water

Of course, bodies cannot live without water. Water does some important things, like carry vitamins and minerals to all the tiniest parts of the body, help food move through the body, keep the body from overheating, and provide some of the minerals bodies need. Some foods, such as melons, cucumber and celery, have a lot of water in them, but aren't eaten in large enough amounts to give all we need. So, drinking water throughout the day is important, especially if playing hard or exercising.

NUTRIENT VARIETY

Foods often have more than one kind of nutrient in them. For example,

- beans have protein, carbohydrate and vitamins. They also have fiber.

- meat has mostly protein but it also has some fat.

- bread is mostly carbohydrate, but also has some protein and vitamins.

- milk has protein and fat in it.

- vegetables and fruits have carbohydrate, vitamins and minerals. They also have fiber.

It's a good idea to eat lots of different kinds of food so you get all the protein, carbohydrate, fat, vitamins and minerals your body needs.

If you learn what type of food is in the things you eat, it will help you to become good at feeding your body wisely.

LET'S DO THIS!
Four Nutrients

To do this activity, you will need

- blank index cards

Steps

1. On each card, write

   a. One of the nutrients,

   b. what your body uses it for,

   c. the foods listed for it.

2. Practice saying the information on the cards until you feel you know it well.

3. Tell another person the information without looking at the cards.

LET'S DO THIS!
Helpers

To do this activity, you will need

- blank index cards

Steps

1. On each card, write down

   a. one of the two helpers,

   b. how each helps your body,

   c. the foods listed for them.

2. Practice saying the information on the cards until you feel you know it well.

3. Tell another person the information without looking at the cards.

CHAPTER 4

Eating Suggestions

CHAPTER 4

Eating Suggestions

It's hard to say exactly what *you* should eat to stay healthy, because not all bodies need exactly the same kinds of foods, and not all families eat the same foods.

But there are some guidelines that will work pretty well for just about everybody.

1. Be sure to eat breakfast.

    If you think about it, at breakfast time your body hasn't had food for around eight hours! It's hungry and needs energy. It needs good food.

    There are a lot of ideas about what makes a good breakfast.

Here are some suggestions for good breakfast foods:

eggs, yogurt, meat or cheese

fruits

vegetables (like in omelets)

hot or cold cereal, made from whole grains and not too much sugar

2. Eat at least two servings of protein each day.

3. Eat lots of fresh vegetables—at least two or three different ones each day.

Raw vegetables have the most vitamins in them. Vegetables that have been cooked have less vitamins, but they are still good for you.

You can get a lot of your vegetables by eating a salad at lunch or dinner every day.

4. Eat some fresh fruit each day.

5. Eat many different kinds of foods. For example, it is probably not a good idea to have mainly peanut butter every day, or only hot dogs for protein, or only fruit.

6. If you eat sweets, like cookies, cake, candy or sodas, eat just a bit, not a lot, and try to save them for the end of the meal.

   Foods like these taste good, but they don't have the vitamins and protein you need for energy, and can put added fat on your body.

These guidelines are a good place to start when you are deciding what to eat to keep your body healthy.

LET'S DO THIS!
Six Suggestions

Steps:

1. Write the six eating suggestions on a piece of paper.

2. Practice saying each in your own words until you can tell another person what they are without reading them.

3. Tell the other person the suggestions in your own words.

LET'S DO THIS!
Create Your Own Meals

To do this activity, you will need

- cookbook

Steps

1. Write down an example of a healthy breakfast. As needed, use a cookbook for ideas on kinds of food to have.

2. Check it against the information in the chapter. Do this at least three times with different foods, or until you are sure you can write down a healthy breakfast.

3. Show it to another person.

4. Repeat steps 1 & 2, with lunches.

5. Show another person.

6. Repeat steps 1 & 2, with dinners.

7. Show another person.

8. Save your written work for a later activity.

LET'S DO THIS!
What Foods Are You Eating?

Steps

1. For the next day, keep a list of all the food you eat at breakfast, lunch and dinner, and also snacks.

2. Next to each one, write which nutrients you think are mostly in them.

3. Then for each meal and snack, say whether you think it was healthy based on the information in the book, and why you think it was or wasn't.

4. Show the list to another person

LET'S DO THIS!
Using Your Meal Plan

Steps

1. Using your written work from the Create Your Own Meals activity on page 27, create a complete meal plan for one day.

2. Show another person your meal plan.

3. Choose one meal from the plan. Work with another person to prepare, or put together, that meal. Enjoy it!

CHAPTER 5

Exercise

CHAPTER 5

Exercise

Exercise means using your body so it stays in good shape.

Running, swimming, riding a bicycle, roller skating, sports—all of these are good for your body. They keep the lungs, heart, muscles, and other parts of the body strong and ready to work when you want them to.

When you exercise, the different parts of your body have to work harder than when you are resting.

Your lungs have to fill with air and empty more often. You call this breathing unless you have to do it fast, and then you call it panting. If your body is in good shape, you can do more exercise or work before you start panting.

Your heart has to pump the blood around your body faster. The blood feeds the different parts of the body and when the body is working harder, it needs to be fed faster, making the heart work faster. If your body is in good shape, the heart is good and strong and can do this extra work well.

Muscles don't grow stronger unless they are used. In fact, if you don't use them, they grow weaker and smaller. When you exercise, you use them and then they can grow larger and stronger.

A body is made to do a lot of things. But if it's kept shut up inside all day, sitting down or lying down all of the time, it can get weak. Then when it's needed for something, it might easily run out of energy.

Be sure you get a lot of exercise. If you want a healthy body—use it!